M

Growth

Written by Joanne Jessop

Wayland

CRISS X CROSS

Picture acknowledgements

The publishers would like to thank the following for allowing their photographs to be reproduced in this book: Associated Press/Topham *frontispiece*; Bruce Coleman Ltd 10 (both/Jane Burton), 11 (Jane Burton), 12 (below/Adrian Davies), 13 (above/Dr Eckart Pott, below Stephen J. Krasemann), 14 (Jan Van de Kam), 16 Peter Davey), 18 (top/Udo Hirsch, centre/Frans Lanting); Francesca Motisi 15 (above); Oxford Scientific Films Ltd 5 (above/main David and Sue Cayless, inset/G. Bernard), 17 (above/G. I. Bernard), 28 (both/Peter Cook), 20 (above/ David Fleetham, below/© Partridge Films Ltd), 21 (Carol Farneti); Papilio Photographic 18 (below/RKP), 22 (both/ADJ), 23 (both/ADJ); Reflections 5 (below/Jennie Woodcock); Tony Stone Worldwide 8 (Patrick Ingrand), 9 (Andy Sacks), 12 (above David Woodfall), 19 (Chris Harvey); Wayland Picture Library 15 (below), 24, 25 (all), 26 (all), 27; ZEFA 4, 6 (both), 7.

Cover photography by Daniel Pangbourne, organized by Zoë Hargreaves.
With thanks to the Fox Primary School.
A special thank you to Ketty and Fabian.

First published in 1993 by
Wayland (Publishers) Ltd
61 Western Road, Hove
East Sussex BN3 1JD, England

© Copyright 1993 Wayland (Publishers) Ltd

Editor: Francesca Motisi
Designer: Jean Wheeler

Consultant: Alison Watkins is an experienced teacher with a special interest in language and reading. She has been a class teacher and the special needs coordinator for a school in Hackney. Alison wrote the notes for parents and teachers and provided the topic web.

British Library Cataloguing in Publication Data
Jessop, Joanne.
Growth. – (Criss Cross)
I. Title II. Series
574.3

ISBN 0-7502-0865-1

Typeset by DJS Fotoset Ltd, Brighton, Sussex
Printed and bound in Italy by L.E.G.O. S.p.A., Vicenza

Contents

Words printed in **bold** in the text are explained in the glossary on page 32.

What is growth?

Growth means getting bigger. All living things start life very small and then grow bigger.

Most young plants and animals look like their parents, but they are smaller. The baby elephant looks like its mother in every way except size.

4

The leaves of this young oak tree are the same shape as the leaves on the parent tree (shown above).

In what ways are these children like their mother?

Plant growth

Plants grow from seeds. When there is enough warmth and water, the seed begins to sprout.

The food stored in the seed passes to the growing roots and stem.

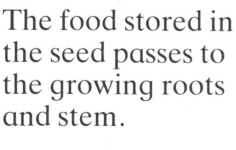

The plant needs
water, light and
warmth so that it
can grow strong
and healthy.

When the plants have used up all the food stored in their seeds they begin to make their own food from sunshine.

Plants need water as well as sunlight. If there is not enough rain you can help plants to grow by using a watering can.

Many plants grow bright flowers. This hibiscus plant grew a bud, which opened up into a flower.

A flower produces seeds from which a new plant can grow. When the flower dies, the seeds are left behind in the ground.

Some plants grow for one season only, then they die. Bluebells start to grow in the spring. They flower, make seeds and die before the summer begins.

Other plants continue to grow all through each year. Trees never stop growing. This oak tree has been growing for nine hundred years.

As each year passes, a tree grows another ring of wood around its trunk. When a tree is cut down you can tell how old it is by counting the rings.

Animal growth

Animals cannot make their own food the way plants do. Animals need to eat food so that they can grow.

Some animals eat other animals. This blue tit is feeding insects to her chicks.

Some animals eat plants. Sheep have to eat grass. Can you think of any other animals that eat grass?

People are animals too. They are able to eat meat as well as plants. Your body needs food and drink so that it will grow strong and healthy.

When an animal has reached its full **adult** size, it stops growing. The length of time it takes to become an adult is not the same for all animals. A young giraffe grows for several years before it reaches adult size.

These baby mice will be fully grown in only a few weeks.

It will be many years before this child is as tall as his father. Most children reach their full height around the age of fifteen.

Non-stop growth

Giant tortoises are different from most animals because they never stop growing!

A baby giant tortoise hatches from an egg. It may live for more than one hundred years, so it will grow very big.

Horses' hooves and rabbits' teeth never stop growing. A vet will clip back this rabbit's teeth.

Some parts of our bodies such as hair and fingernails never stop growing.

How big will it grow?

Some living things grow very big. Whales are the biggest animals in the sea. These are humpback whales.

The biggest trees are the giant redwoods of California, USA. ▶

Ostriches are the biggest birds in the world. They are unable to fly because they are so big and heavy.

Metamorphosis

Some animals change from one form into another as they grow. This is called **metamorphosis**.

When a caterpillar is fully grown its skin becomes a hard, crusty shell called a pupa. Inside the pupa the caterpillar changes and grows into a butterfly.

Then the butterfly
hatches out of the
pupa and flies away.

Growing older

As a person grows older many changes take place.

A mother gives birth to a baby. The baby learns to walk and talk and grows into a child.

Children grow taller and learn more as they grow older.

When children grow into teenagers they have usually reached their adult height.

A teenager grows into an adult. As the years pass, the adult grows older in age, but no longer grows in height.

26

Look at these pictures of people of different ages. What changes can you see? What stays the same as people grow older?

Making things bigger

Plants and animals are not the only things that grow. Non-living things grow when more **material** is added to them.

This weaver bird is building a nest.

The nest grows bigger as the bird adds more grass to it.

These balloons
are getting
bigger as the
children blow
air into them.

You can make a
tower grow bigger
by piling blocks on
top of each other.

Notes for parents and teachers

Maths
- Focus on the mathematical language of growth and size e.g. tall, taller, small, smaller, high, higher, highest, etc.
- Investigate how different parts of the children's bodies have grown. How do they know? Chart results using skills of observation, comparing and classifying with some challenge in measuring (standard and non-standard units).
- Draw around the children, cut out the figure and group according to different criteria.
- Gather information about average weight/length of babies. Compare with their weight/height now. Ask parents or friends to bring in a baby or toddler to help comparisons.
- Compile statistics about themselves and make a zig-zag book. Consider food and growth, weight, measurements, how much sleep we have, etc.
- Explore time – weeks, months, years. For instance the children could work out the number of birthdays they have had.

History
- Use the children, parents and grandparents as primary sources.
- Talk about baby clothes and how small they are. Discuss what the children could and couldn't do when they were babies. Record through drawing, writing, and wall stories.
- Children can complete a personal questionnaire or log book recording their changes since birth.

Dance
- Individuals, groups or the whole class can be used as something that is growing or shrinking. Use poetry, music, art work, sounds etc as a stimulus.

Design Technology
- Making things fit, e.g. shoe size – biggest and smallest feet. Children could design and make their own pair of slippers.

Science
- Growing things – increase in size and weight.
- Things look different when fully grown.
- Human beings are constantly changing.
- Plants and humans are living things which need certain conditions for growth and development.

Language
- Make individual, group or class books. A possible theme could be "Our journey from baby to old person".
- Make up a rhyme or jingle to read and re-read, e.g. "Who helped me to grow?" or "What made me grow?"
- "What if?" language games. For example "What if your hands grew as big as a table?" Use the ideas as a basis for discussion, art work, story-writing, taping or drama.
- Fiction books with growth as a theme and also a favourite with young children.
  Titch and *You'll soon Grow into them Titch* (Pat Hutchins 1985 Picture Puffin)
  The Enormous Turnip (Mary Shepherd 1989 Collins)
  Janine and the New Baby I. Thomas 1991 Little Mammoth)
  The Shrinking of Treehorn (F. Parry Heide Kestrel/Young Puffin)
  Alice in Wonderland (L. Carroll)
  Jack and the Beanstalk
  also try the modern version called *Jim and the Beanstalk*

Health Education
- How do people know I am growing? What new things can I do? Where do I go? What can I reach? Use magazines and cut out a variety of people of all ages doing different activities. Sort into sets. (Take the opportunity to discuss the varied and non-stereotypical roles of men and women, young and old.)
- Contact the local Health Education Authority for books, resources and teaching materials.

Underlying concepts which can be covered:
- Self-image • Growth • Change • Time
- Respect for other ways of being and living.
- Growing things indoors:

cress and mustard	5-7 days
fruit pips	2-3 weeks
bird seed or hamster food	1-5 weeks
bulbs	6-8 weeks

- Looking after pets (fish, hamsters, gerbils, tadpoles) will not only give children a sense of responsibility but also an understanding of how different animals carry out their lives, breed, rear their young, grow and die.

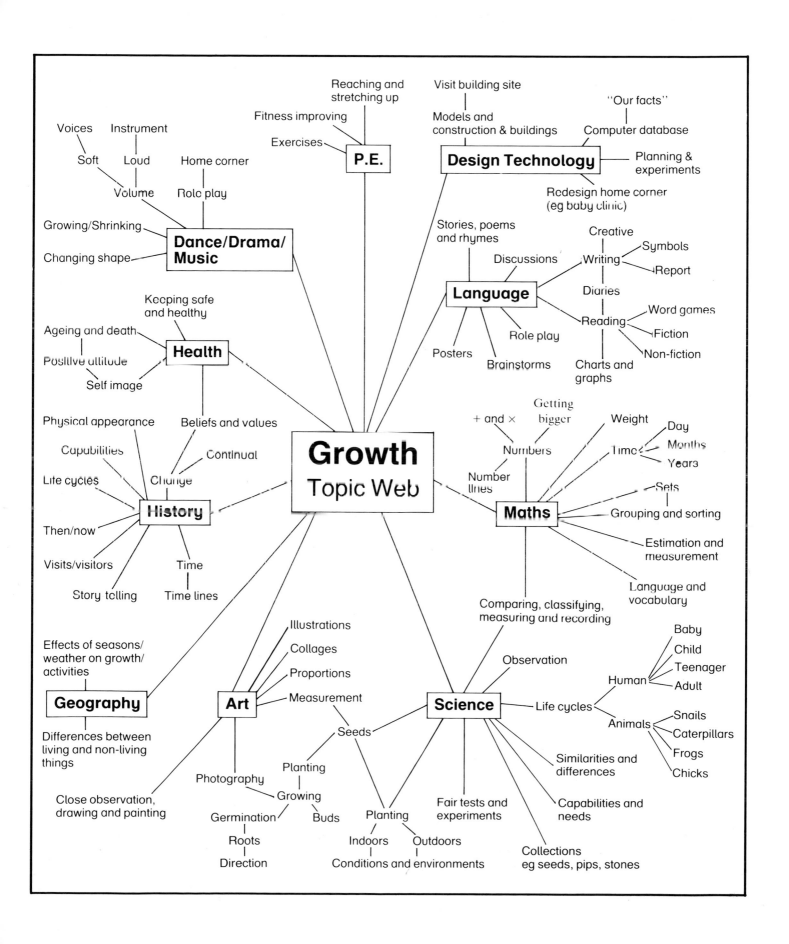

Glossary

Adult   Grown up.

Metamorphosis   The changes in form that certain animals go through as they grow and develop.

Material   What a thing is made of.

Index